Bibliographic information published by the German National Library:

The German National Library lists this publication in the National Bibliography; detailed bibliographic data are available on the Internet at http://dnb.dnb.de .

Imprint:

Copyright © 2018 GRIN Verlag
Print and binding: Books on Demand GmbH, Norderstedt Germany
ISBN: 9783668820203

This book at GRIN:

https://www.grin.com/document/442783

Gibson Lewa

The Osmosis of Potato Strips

GRIN Verlag

Aim

To investigate the change in mass in potato strips over a period of two hours when immersed in distilled water (hypotonic solution) and salty water (hypertonic solution).

Research Question

How does the size of potato strips when immersed in both distilled water and salty water change over a period of 2 and half hours measured at 30 minutes intervals?

Background Information

Osmosis is one of the physiological processes in living organisms, among them active transport and diffusion. Osmosis is the movement of water molecules from a region of low concentration to a region of high concentration across the semi-permeable membrane. Os miss is impo4ettnt in both plants and animals. In plants it makes cells to be turgid while in animals it offsets the osmotic pressures in the cell. Plant cells are hypertonic because they have a cell sap, so when they are pout in distilled water (hypotonic solution), it absorbs water by osmosis, swells up and become turgid. They do not burst because they have a cell wall that develops a wall pressure that balances the turgor pressure exerted by turgid cells. As the plant gains turgidity, its volume increases until it achieves maximum turgidity, water will then start moving out of the cell to balance the pressure in the cells and outside environment.

When a plant cell is placed in the hypertonic solution for example solution containing NaCl $_{(aq)}$, it loses water due to osmosis, shrinks and become flaccid (plasmolysis). This is because the salt concentration is high then the concentration in the cell sap.

It is important to know how plant cells when placed in different solutions first to satisfy curiosity and understand water relations in plants that can be applied in studying various physiological processes in fields such as plant biochemistry and pathology among others.

Variables

Type of variable	Variables	Unit	Range
Dependent	The change in potato size	Mass (g)	
Independent	Time taken for osmosis to occur (to observe a change in potato cell)	Minutes	30, 60, 90, 120, 150

Dependent: the change in potatoes depends on the molarity (concentration) of salt solution. As the salt concentration increases, there is a decrease in mass of potato tissue.

Independent variable: time here is an independent variable; its change is not affected by either mass of potato tissue or the change in salt concentration. Will be measured at 30 minutes intervals and the dependent variable (mass of potato) will be got at that time.

Mass of potato tissue is calculated as;

Change in Mass = Initial Mass - Final Mass

Initial mass is the mass of fresh potato before adding any salt solution while the final mass is the mass of potato after adding salt solution at a specific time interval.

Controlled Variables	How it is Controlled
The salt concentration solution	All five Potatoes of different masses were placed in different beakers each containing salt of the same concentration for equal time intervals
The size and shape of the potato	Potatoes were of different sizes, and therefore a change was recorded, and the percentage change in mass was calculated based on the total number of potatoes.
Source of potatoes	All potatoes were obtained from the same bag, had the same color and same shape but different sizes.
How potato tissues are immersed inside the salty water	All potatoes were immersed fully in salty water with the same molarity. The potatoes were changed as the molarity of the solution was also increased.
the temperature needed for the experiments	All the experiments were carried out under normal standards conditions of temperature and pressure.
Contact with external enlivenment	The whole setup was in contact with the external environment as the physiological process was taking place.

For safety reasons, there was no need of gloves since the reagents used and the entire requirements are not harmful. However, the knife was held firmly while using it to avoid cuts.

Care was observed when handling glass apparatus since they are fragile and can easily when they slide over the working bench.

Materials and Apparatus

Materials	
Material	Description
Water	500ml
Potato	20 pieces
Salt	56.671875g

Apparatus	
Name	Quantity
Stopwatch	1
weighing scale	1
cutting board	1
knife	1
beaker	5
salt $NaCl_{(s)}$	1
ruler	1
spoon	1
Small beaker (100 ml)	5

Procedure

Potatoes were measured using a weighing balance and their masses recorded. 100 ml of distilled water was measured and put it in five different beakers. The first beaker was not added any NaCl (aq), second, third and fourth beakers were added different masses of $NaCl_{(aq)}$ as follows respectively, 3.656g, 7.311g, 14.625g and 29.25g and the salt were allowed to dissolve completely. The measure potatoes were immersed in each of the beakers containing different concentrations of $NaCl_{(aq)}$ solution at the same time. Their masses were recorded for each of the beakers after every 30 minutes for two and half hours.

Results

Table 1: Mass of potato strips at 30 minutes intervals in salty water 0f 0.0 molarity

Potatoes	Mass of potato before putting in water with salt 0.12 molarity	Mass of potato strips at 30 minutes intervals in salty water 0f 0.12 molarity				
		After 30 min water	After 1 hour in water	After 1.3 hours in water	After 2 hours in water	After 2.30 hours in water
Potato 1	12.384g	12.280g	12.608g	14.103g	13.115g	13.160g
Potato 2	14.728g	13.993g	14.093g	12.699g	15.048g	15.122g
potato 3	13.804g	14.922g	11.402g	15.069g	11.510g	14.143g
potato 4	11.215g	12.539g	14.960g	11.462g	14.120g	12.714g
potato 5	12.749g	12.915g	12.976g	13.075g	12.703g	11.494g
Mass of all potatoes	64.88g	66.649g	66.039g	66.408g	66.496g	66.633g

Table 2: Mass of potato strips at 30 minutes intervals in salty water 0f 0.12 molarity

Potatoes	Mass of potato before putting in water with no salt	Mass of potato strips at 30 minutes intervals in salty water 0f 0.0 molarity				
		Mass of potato after being inside water 30 min	Mass after 1 hour	Mass after 1.30	Mass after 2h	Mass after 2.3h
Potato 1	14.609g	15.504g	12.472g	13.805g	13.843g	13.844g
Potato 2	16.147g	12.115g	10.449g	15.947g	15.966g	16.022g
Potato 3	11.089g	17.179g	15.860g	12.541g	17.802g	17.835g
Potato 4	12.379g	13.408g	17.654g	17.752g	12.612g	12.662g
Potato 5	9.386g	10.121g	13.702g	10.512g	10.562g	10.664g
All potatoes	63.61g	68.227g	70.137g	70.557g	70.785g	71.027g

Table 3: Mass of potato strips at 30 minutes intervals in salty water 0f 0.25 molarity

Potatoes	Mass of potato before putting in water with salt 7.311g	Mass of potato strips at 30 minutes intervals in salty water 0f 0.25 molarity				
		Mass after 30min in water	Mass after 1h in water	Mass after 1.3hour in water	Mass after 2h in water	Mass after2.3 hours in water
Potato 1	13.501g	11.570g	11.100g	12.540g	12.398g	10.830g
Potato 2	12.682g	13.785g	12.737g	13.192g	10.936g	15.757g
Potato 3	12.140g	13.026g	16.466g	11.044g	13.037g	12.312g
Potato 4	14.618g	16.785g	13.490g	10.809g	15.946g	12.937g
Potato 5	12.208g	11.432g	11.270g	16.129g	10.642g	10.535g
Mass of all potatoes	70.149g	66.598g	65.063g	63.714g	62.959g	62.371g

Table 4: Mass of potato strips at 30 minutes intervals in salty water

Potatoes	Mass of potato before putting in water with salt 14.625g	Mass of potato strips at 30 minutes intervals in the salty water of 0.5 molarity				
		After 30min in water	After 1h in water	After 1.3h in water	After 2h in water	After 2.3h in water
Potato 1	12.883g	10.183g	7.588g	9.990g	9.656g	9.450g
Potato 2	11.143g	11.203g	9.043g	10.417g	8.333g	8.858g
Potato 3	11.631g	7.990g	10.943g	9.310g	9.020g	8.147g
Potato 4	8.938g	11.591g	10.484g	8.622g	7.105g	6.962g
Potato 5	12.984g	9.671g	9.636g	7.275g	9.990g	9.779g
Mass of all potatoes	57.579g	50.638g	47.694g	45.614g	44.104g	43.196g

Table 5: Mass of potato strips at 30 minutes intervals in salty water 0f 1.0 molarity

Data Analysis

Potatoes	Mass of potato strips at 30 minutes intervals in salty water 0f 1.0 molarity					
	Mass of potato before putting in water with salt 29.25g	After 30min in water	After 1h in water	After 1.3h in water	After 2h in water	After 2.3h in water
Potato 1	12.740g	10.816g	10.509g	11.722g	11.462g	12.824g
Potato 2	14.314g	12.304g	10.346g	13.281g	7.179g	9.785g
Potato 3	17.026g	10.760g	11.884g	7.340g	9.866g	11.364g
Potato 4	9.117g	14.052g	7.503g	10.288g	10.076g	7.134g
Potato 5	12.372g	7.738g	13.526g	10.081g	12.939g	9.940g
Mass of all potato	65.569g	55.67g	53.768g	52.712g	51.522g	51.047g

Table 6: Average change in mass at 30 minutes intervals and the overall change in mass after 2 hours 30 minutes

Molarities of salt solution	The average mass of all pota-toes	The average change in mass in					Parentage change in mass (total)
		30 min	60 min	90 min	120 min	150 min	11.660%
0.0	63.61g	68.227g	70.137g	70.557g	70.785g	71.027g	
0.12	64.88g	66.649g	66.039g	66.408g	66.496g	66.633g	2.702%
0.25	70.149g	66.598g	65.063g	63.714g	62.959g	62.371g	-11.088%
0.5	57.579g	50.638g	47.694g	45.614g	44.104g	43.196g	-27.980%
1.0	65.569g	55.67g	53.768g	52.712g	51.522g	51.047g	-21.423%

Table 7: Initial and Final Mass of Potato Tissue

Time in Minutes	Initial and Final Mass of Potato Tissue									
	Molarity									
	0.0		0.12		0.25		0.5		1.0	
	Initial	Final	Initial	Final	Initial	Final	Initial	Final	Initial	Final
30	63.61g	68.227g	64.88g	66.649g	70.149g	66.598g	57.579g	50.638g	65.569g	55.67g
60	63.61g	70.137g	64.88g	66.039g	70.149g	65.063g	57.579g	47.694g	65.569g	53.768g
90	63.61g	70.557g	64.88g	66.408g	70.149g	63.714g	57.579g	45.614g	65.569g	52.712g
120	63.61g	70.785g	64.88g	66.496g	70.149g	62.959g	57.579g	44.104g	65.569g	51.522g
150	63.61g	71.027g	64.88g	66.633g	70.149g	62.371g	57.579g	43.196g	65.569g	51.047g

Table 8: Mean, SD and 33.33% of the mean

Duration for osmosis to occur in minutes	Mean, SD and 33.33% of the mean		
	Mean	SD	33.33% of the mean
30	62.96	6.25	20.98
60	62.45	7.24	20.81
90	62.08	7.36	20.69
120	61.77	8.42	20.58
150	61.60	8.72	20.53

As all Standard deviation values are less than 33.33% of the mean and hence can be seen as accurate values.

Graph 1: Mean mass of potatoes with standard deviation

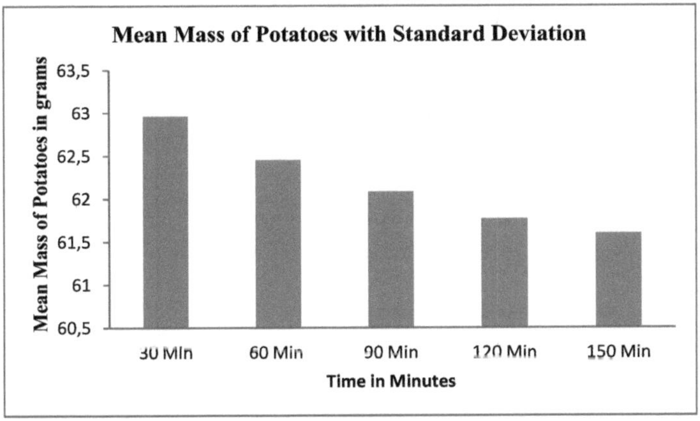

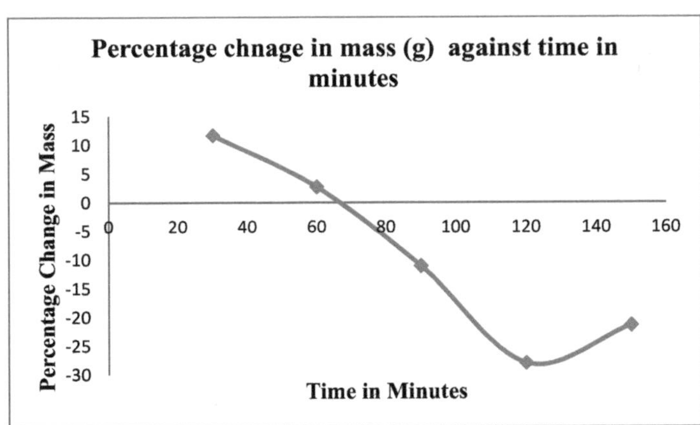

The mass of potato tissues generally reduces with time for all experiments except one with distilled water

Graph 2: Percentage change in mass (g) against time in minutes

Graph 3: Molarities of salt concentration against change in mass

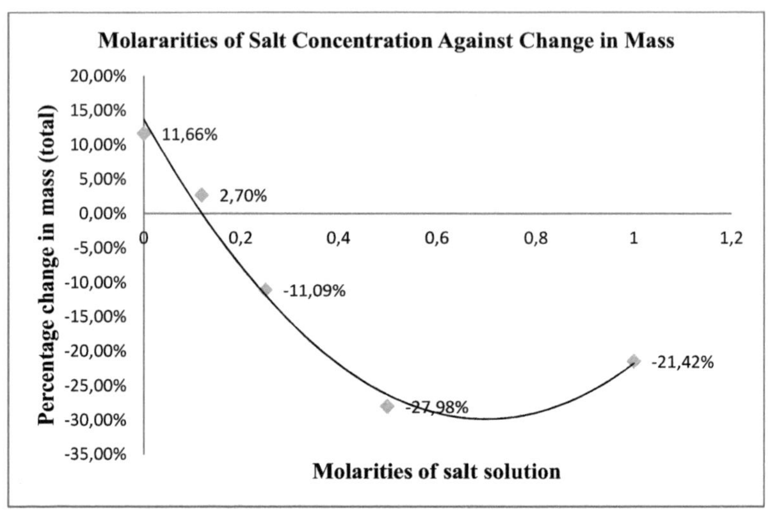

As the salt concentration is increased, there is a large reduction in mass.

Calculations Involved and other Adjustments

Average Mass is obtained by adding masses of;

$$\frac{Potato\ 1 + Potato\ 2 + Potato\ 3 + Potato\ 4 + Potato\ 5}{5}$$

Change in mass is obtained by;

Final Mass - Initial Mass

Percentage change in mass is obtained by; $\dfrac{\text{Average Mass at 150 Min} - \text{Average Initial Mass}}{\text{Average Initial Mass}} \times 100$

Standard deviation is calculated using the formula;

$$\sigma = \sqrt{\frac{1}{N}\sum_{i=1}^{N}(x_i - \overline{x})^2}.$$

Where $\bar{x}$ is the mean,

X_l is each individual mass ,

N is the number of values,

Σ is Summation and

σ is Standard deviation.

Standard deviation was calculated from each individual values from Table 7 as follows,

i.   Work out the mean of all numbers in the range

ii.   Then for each number: subtract the Mean and square the result

iii.   Then work out the mean of those squared differences.

iv.   Finally, take the square root of the differences

33.33% of the mean is calculated as;

$$\frac{33.33}{100} \times \text{the mean mass of potato}$$

The data obtained is nearly accurate because the errors are minor. The areas will be further discussed at evolution (at the end).

Measurement of the volume of water was nearly accurate using beakers. The concentrations of solutions were nearly accurate also due to using of electronic balance though there are human errors which cannot be assumed.

Interpretation

From the results, it is clear that as the concentration of salt solution increases, there is a larger decrease in the size of the potato. For example, at 0.0 concentration, that is, distilled water, percentage change is 11.660%, at 0.12 molarity, change is 2.702%, at 0.25 molarity, change is -11.088%, at 0.5 molarity, change is-27.980%, while at 1.0 molarity, change is -21.423%.

Discussion

In osmosis, water which is uncharged molecules passes through the cell membrane. A cell membrane is a semi-permeable membrane, that is, it only allows small particles to go through it. Osmosis depends on its concentration gradient, from a region of low concentration as it moves to a region of high concentration. It is dependent on temperature, size of the molecule, thinness of the membrane and the concentration gradient. In this activity, osmosis in potato cell was studied. Salty water with $Cl^-_{(aq)}$ is concentrated as compared to the cell sap, that is, hypertonic solution. The cell sap, therefore, loses water by osmosis and shrink. From the results, as the concentration of salts increases, the mass of potato reduces, therefore solution with 0.0 molarity of NaCl (aq) recorded an increase in mass while the ones with 1.0 molarity of NaCl (aq) recorded a great decrease in mass. The process where a plane cell loses water, shrink and become flaccid is called deplasmolysis. A plasmolyzed cell can be made turgid by immersing it in distilled water for some time (deplasmolysis) (Odom et al, 2017).

Therefore for osmosis to occur, the two solutions must be separated by a semi-permeable, membrane, for this case, the potato cell membrane is only permeable to water molecules from potato cell but is not permeable to salt molecules from $Cl^-_{(aq)}$ ions that are present in water solution. At 30 min, all of them showed a decrease in mass because water is drawn from potato tissue to the salt solution. As the time increases, the cell is continually losing water; to offset the osmotic imbalance. The overall change in mass is a decrease, in the end, the solutions become isotonic, that is, all solution s have attained the same concentration, and there is no net movement of water molecules from one solution to another.

The main aim of osmosis is to offset osmotic imbalances in bodies of living organisms to provide optimum conditions for body functions. Osmosis continues to take place until when the two solutions are almost equal in concentration.

As osmosis is taking place, a force will be created in the hypertonic solution that aims to prevent osmosis from taking place; such a force is called osmotic pressure. If osmotic pressure of a solution is high, it will draw a lot of water from the adjacent solution, and if it is low, it will draw little waiter. Osmotic pressure is, therefore, a measure of the total amount of dissolved salts in water (Xu et al, 2017).

Conclusion

As is can be seen from Table 6, there is generally a decrease in mass when a potato is placed in water containing NaCl (aq) solution. The potato sap has little solutes, and therefore it is hypotonic while the salt solution has more solutes. Therefore, it is hypertonic. Water molecules moved from a region of low concentration to a region of high concentration. The purpose of using five potato tissues is for the accuracy of the results obtained. The percentage change in mass calculated increased the increase in salt concentration. At 30 min, all of them showed a decrease in mass. This is due tote fact that a lot of water is drawn from potato tissue to the salt solution. As the time increases, the cell is continually losing water; there is no general trend as in others it increases from 30 minutes to 60 minutes while in others it decreases. Near to the end of the experiment, the cell increased slightly in mass, though it is still lower than the initial value. This is because the solution is isotonic, that is, all solution s have attained the same concentration, and there is no net movement of water molecules from one solution to another.

Though there is no specific theoretical value about the mass of potato used, the general conclusion is of a decrease in mass as the cells are losing water by osmosis (Odom et al, 2017).

Evaluation

Limitations and Weaknesses	How it Affected the Outcome of the Experiment	How to Improve
Human error	Though there is a need to be very accurate, there usually some small errors such as measured using a stopwatch for time, measuring volume using a measuring cylinder, etc. This affected the outcome as there is no uniformity among the appeasers being studied.	Ensuring that all potatoes are put into the beaker all within one minute to enhance accuracy. Reading the volume on the measuring cylinder down the meniscus to avoid parallax errors.
The temperature of the salt solution was not measured	Salt solutions can be of different temperatures depending on the concentration and maybe while stirring to dissolve. Osmosis takes place very fast at high temperature. There is a need to check the	Use a thermometer to check the temperature and also to, maintain it within a certain range.

	temperature of salt solutions to arrive at amicable conclusions.	
Time intervals to collect data	The data was measured at 30minutes intervals for two hours. Though it is more accurate, there are data gaps between 1 minute and 30 minutes. This might lead to a vague conclusion.	Record data after every ten minutes for even up to four hours to reduce data gaps and enhance accuracy.

Reference List

Odom, A.L., Barrow, L.H. and Romine, W.L., 2017. Teaching Osmosis to Biology Students. *The*

American Biology Teacher, 79(6), pp.473-479.

Xu, W., Chen, Q. and Ge, Q., 2017. Recent advances in forward osmosis (FO) membrane:

Chemical modifications on membranes for FO processes. *Desalination, 419*, pp.101-116.

YOUR KNOWLEDGE HAS VALUE

- We will publish your bachelor's and master's thesis, essays and papers

- Your own eBook and book - sold worldwide in all relevant shops

- Earn money with each sale

Upload your text at www.GRIN.com and publish for free